SPECIFICATION WRITERS' GUIDE
For Federal Lands Highway

Publication No. FHWA-CFL/TD-08-001   May 2008

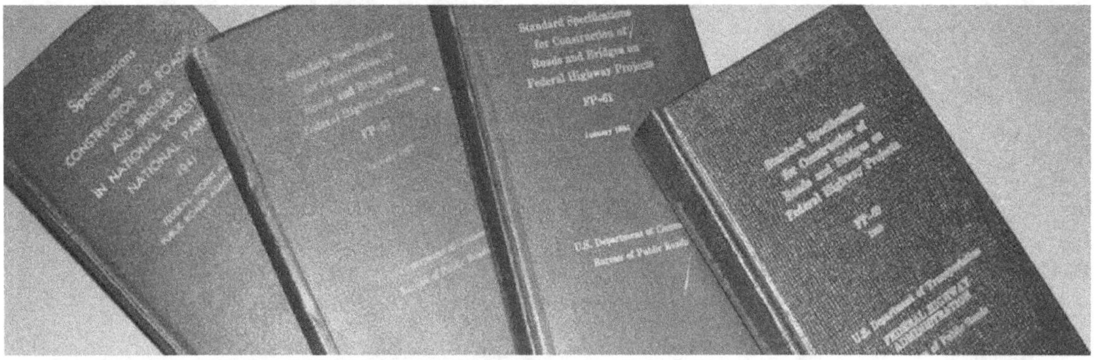

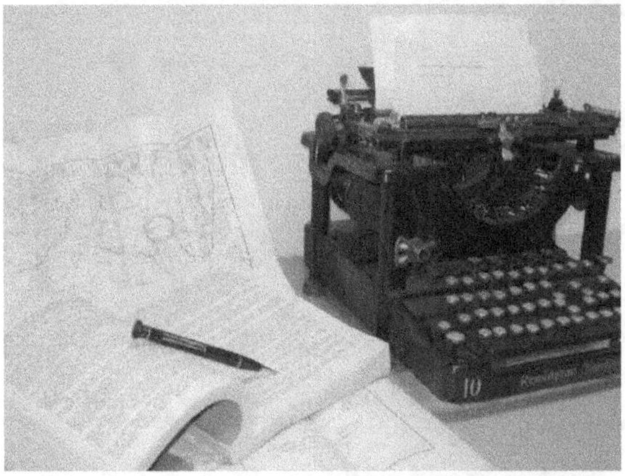

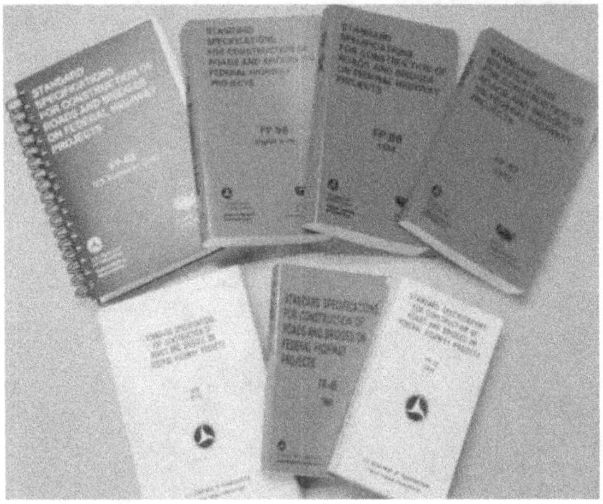

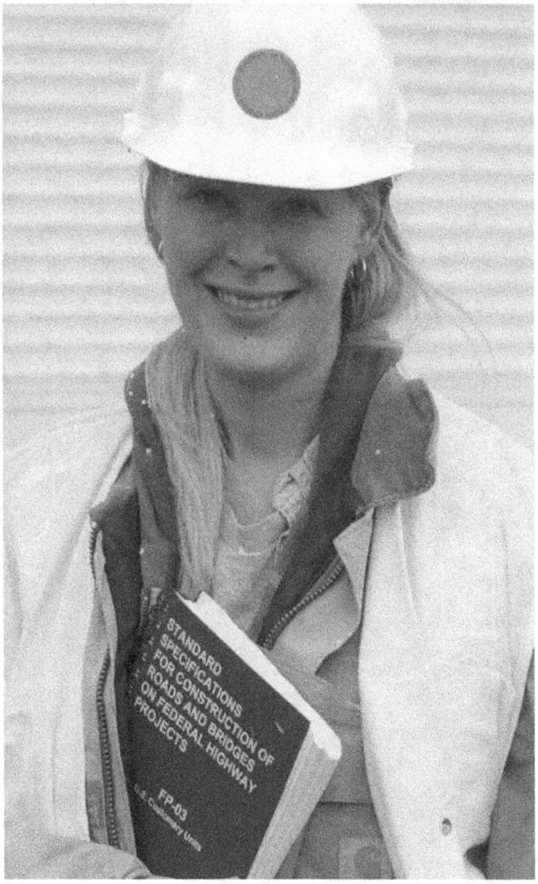

U.S. Department
of Transportation

Federal Highway
Administration

FORWARD

This Guide is intended for writers that develop specifications for the Federal Highway Administration's (FHWA) Federal Lands Highway Program (FLH). This Guide is to promote and facilitate the writing of specifications that conform to the five Cs of specification writing -- clear, concise, complete, correct, and consistent. Adherence to the guidance provided in these chapters and Chapter 9.4.11 of the FLH's Project Development and Design Manual will help writers develop well-written specifications for use with the *Standard Specification for Construction of Roads and Bridges on Federal Highway Projects (FP)*. The FP in and of itself is a good example of proper style and format.

For questions not addressed in these documents refer to *The Chicago Manual of Style* or to a standard English dictionary.

F. David Zanetell, P.E., Director of Project Delivery
Federal Highway Administration
Central Federal Lands Highway Division

Notice

This document is disseminated under the sponsorship of the U.S. Department of Transportation in the interest of information exchange. The U.S. Government assumes no liability for the use of the information contained in this document. This report does not constitute a standard, specification, or regulation.

The U.S. Government does not endorse products or manufacturers. Trademarks or manufacturers' names appear in this report only because they are considered essential to the objective of the document.

Quality Assurance Statement

The FHWA provides high-quality information to serve Government, industry, and the public in a manner that promotes public understanding. Standards and policies are used to ensure and maximize the quality, objectivity, utility, and integrity of its information. FHWA periodically reviews quality issues and adjusts its programs and processes to ensure continuous quality improvement.

Technical Report Documentation Page

1. Report No. FHWA-CFL/TD-08-001	2. Government Accession No.	3. Recipient's Catalog No.

4. Title and Subtitle *Specification Writers' Guide* *For Federal Lands Highway*	5. Report Date May 2008
	6. Performing Organization Code

7. Author(s) Linda Konrath, Sydney Scott P.E., and Lauralee Marsiglio	8. Performing Organization Report No.

9. Performing Organization Name and Address Trauner Consulting Services, Inc. 1617 JFK Boulevard, Suite 600 Philadelphia, PA 19103	10. Work Unit No. (TRAIS)
	11. Contract or Grant No. DTFH68-06-P-00182

12. Sponsoring Agency Name and Address Federal Highway Administration Central Federal Lands Highway Division 12300 W. Dakota Avenue, Suite 210 Lakewood, CO 80228	13. Type of Report and Period Covered Final Report: Guidelines August 2006 – April 2008
	14. Sponsoring Agency Code HFTS-16.4

15. Supplementary Notes

COTR: Roger Surdahl, FHWA-CFLHD. *FHWA-FLH Specification Coordination* Group: David Green, FHWA-FLH, Wade Johnson, FHWA-WFLHD, Michael Peabody, FHWA-CFLHD, Steve Arnold, FHWA-EFLHD. *Advisory Panel Members:* Greg Dolson, John Seabrook, Mark Taylor, and Karen Pinell, FHWA-FLH; Ron Andresen and Shari Brandt, FHWA-CFLHD; Gary Brown, Mark Clabaugh, Rajan Patel, FHWA-EFLHD; Amit Armstrong, Brent Coe, Marty Flores, Brian Minor, and John Snyder, FHWA-WFLHD; Bernie Kuta, FHWA-RC; and Ken Jacoby, FHWA-HQ. This project was funded under the FHWA Federal Lands Highway Technology Deployment Initiatives and Partnership Program (TDIPP).

16. Abstract

This document contains guidelines to help writers develop specifications for the Federal Highway Administration's Federal Lands Highway program. Topics addressed include:
- Specification writing style;
- Organization and format,
- Proper terminology and phrasing;
- Capitalization and abbreviation; and
- Punctuation and grammar rules.

17. Key Words **SPECIFICATIONS, STYLE GUIDE, WRITING, DEVELOPMENT OF SPECIFICATIONS, ACTIVE VOICE, IMPERATIVE MOOD**	18. Distribution Statement No restriction. This document is available to the public from the sponsoring agency at the website http://www.cflhd.gov.

19. Security Classif. (of this report) Unclassified	20. Security Classif. (of this page) Unclassified	21. No. of Pages 44	22. Price

Form DOT F 1700.7 (8-72)　　　　　　　　　　**Reproduction of completed page authorized**

SI* (MODERN METRIC) CONVERSION FACTORS

APPROXIMATE CONVERSIONS TO SI UNITS

Symbol	When You Know	Multiply By	To Find	Symbol
LENGTH				
in	inches	25.4	Millimeters	mm
ft	feet	0.305	Meters	m
yd	yards	0.914	Meters	m
mi	miles	1.61	Kilometers	km
AREA				
in^2	square inches	645.2	square millimeters	mm^2
ft^2	square feet	0.093	square meters	m^2
yd^2	square yard	0.836	square meters	m^2
ac	acres	0.405	Hectares	ha
mi^2	square miles	2.59	square kilometers	km^2
VOLUME				
fl oz	fluid ounces	29.57	Milliliters	mL
gal	gallons	3.785	Liters	L
ft^3	cubic feet	0.028	cubic meters	m^3
yd^3	cubic yards	0.765	cubic meters	m^3
NOTE: volumes greater than 1000 L shall be shown in m^3				
MASS				
oz	ounces	28.35	Grams	g
lb	pounds	0.454	Kilograms	kg
T	short tons (2000 lb)	0.907	megagrams (or "metric ton")	Mg (or "t")
TEMPERATURE (exact degrees)				
°F	Fahrenheit	5 (F-32)/9 or (F-32)/1.8	Celsius	°C
ILLUMINATION				
fc	foot-candles	10.76	Lux	lx
fl	foot-Lamberts	3.426	$candela/m^2$	cd/m^2
FORCE and PRESSURE or STRESS				
lbf	poundforce	4.45	Newtons	N
lbf/in^2	poundforce per square inch	6.89	Kilopascals	kPa

APPROXIMATE CONVERSIONS FROM SI UNITS

Symbol	When You Know	Multiply By	To Find	Symbol
LENGTH				
mm	millimeters	0.039	Inches	in
m	meters	3.28	Feet	ft
m	meters	1.09	Yards	yd
km	kilometers	0.621	Miles	mi
AREA				
mm^2	square millimeters	0.0016	square inches	in^2
m^2	square meters	10.764	square feet	ft^2
m^2	square meters	1.195	square yards	yd^2
ha	hectares	2.47	Acres	ac
km^2	square kilometers	0.386	square miles	mi^2
VOLUME				
mL	milliliters	0.034	fluid ounces	fl oz
L	liters	0.264	Gallons	gal
m^3	cubic meters	35.314	cubic feet	ft^3
m^3	cubic meters	1.307	cubic yards	yd^3
MASS				
g	grams	0.035	Ounces	oz
kg	kilograms	2.202	Pounds	lb
Mg (or "t")	megagrams (or "metric ton")	1.103	short tons (2000 lb)	T
TEMPERATURE (exact degrees)				
°C	Celsius	1.8C+32	Fahrenheit	°F
ILLUMINATION				
lx	lux	0.0929	foot-candles	fc
cd/m^2	$candela/m^2$	0.2919	foot-Lamberts	fl
FORCE and PRESSURE or STRESS				
N	newtons	0.225	Poundforce	lbf
kPa	kilopascals	0.145	poundforce per square inch	lbf/in^2

*SI is the symbol for the International System of Units. Appropriate rounding should be made to comply with Section 4 of ASTM E380.
(Revised March 2003)

TABLE OF CONTENTS

LIST OF TABLES

LIST OF FIGURES

COVER PHOTOS: *(Top) FP Specifications for 1941, 1957, 1961, and 1969. (Lower Left) FP Specifications for 1974, 1979, 1985, 1992, 1996, and 2003. (Lower Right) Carolyn Burg.*

CHAPTER 1 — SPECIFICATION DEVELOPMENT

1.1   GENERAL

There are two basic specification types that are developed by writers in Federal Lands Highway (FLH): Unique Project Specifications to address a unique requirement for a single project and Supplemental Specifications (SS) to be approved for standard use on specified projects. Both types of specification are used in the Special Contract Requirements (SCRs) for FLH construction projects. For more information, see the FLH Project Development and Design Manual (PDDM) Chapter 9.6.9.

1.2   INSTRUCTIONS TO READERS

All specifications written either add, delete, or amend the requirements of the FP. In order to convey the intent of the revision to the reader, instructions must be given. The following are some common instructions for adding, deleting, and amending requirements of the FP:

(Added Subsection).
Add the following:
Add the following after paragraph (2):
Add the following to the material list:
Delete Section XXX and substitute the following:
Delete this Subsection and substitute the following:
Delete paragraph (2) and substitute the following:
Delete the text paragraph (a) and substitute the following:
Delete Table ###-# and substitute the following:
Amend as follows:

> *Example*:
>
> **718.08(b)(2)***(c)* Delete this paragraph and substitute the following:
>
> | *(c)* Galvanizing after punching | ASTM A 653M |
> | (inside and outside of post) | coating Z275 designation |
>
> **107.10 Environmental Protection.** Add the following:
>
> Do not fuel equipment with in 150 feet of waterways.

Use *"Amend as follows"* when more than one instruction is used in a Subsection.

1.3   SUBMITTING PROPOSED SPECIFICATIONS FOR CONSIDERATION

Proposed specification changes should first be submitted through Division's Specification Team. Proposed Supplemental Specification to change the FP may then be submitted to any member of the FLH Specification Coordination Group (SCG). In addition to the proposed specification and instruction to readers, a statement as to when the specification is to be used should be included.

Example:

Include the following in projects where Superpave hot asphalt concrete pavement design is required.

Also provide a statement describing why the revision is necessary and include any background information as appropriate.

Example:

REASON: At the May 2005 MTT meeting it was decided that the specifications for rotational viscosity and mass loss should be deleted and the specifications for bending beam rheometer and direct tension changed.

See Appendix A for a sample proposed specification submittal.

CHAPTER 2 — ORGANIZATION OF SPECIFICATIONS

Specifications are divided into various Divisions and Sections (see PDDM Subchapter 9.6.9.8).

Each section uses the hierarchy shown below to distinguish varying levels of information. Use only as many sublevels as necessary to clearly organize and convey specification requirements. Add content-specific headings where possible.

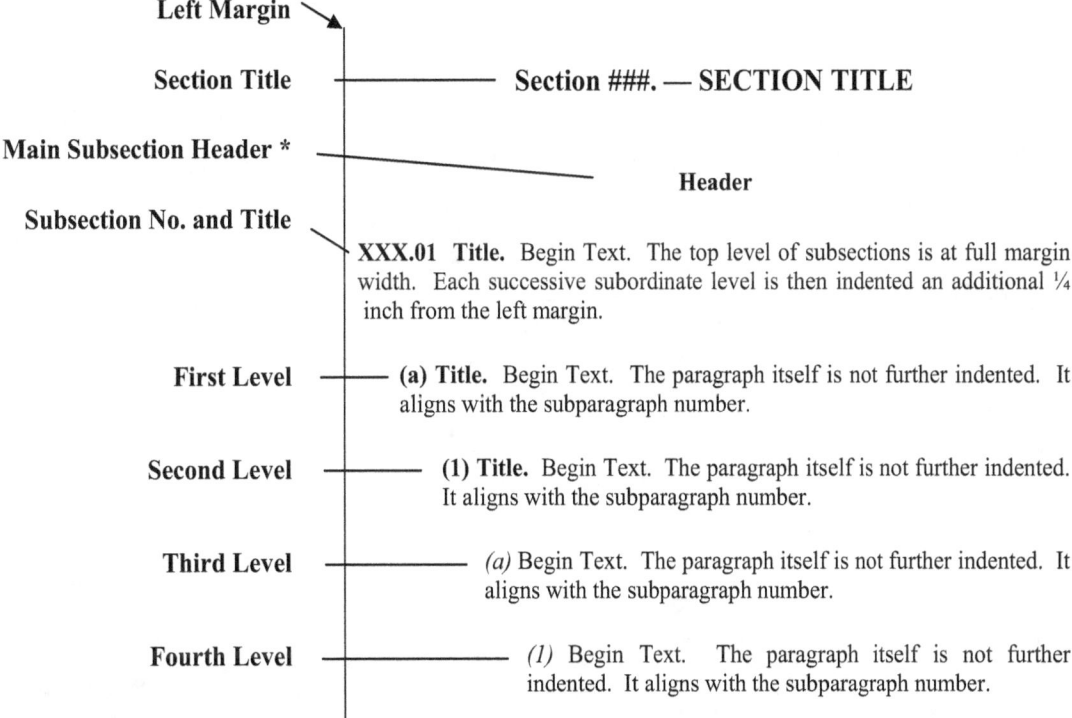

Left Margin

Section Title — Section ###. — SECTION TITLE

Main Subsection Header * — Header

Subsection No. and Title

XXX.01 Title. Begin Text. The top level of subsections is at full margin width. Each successive subordinate level is then indented an additional ¼ inch from the left margin.

First Level — **(a) Title.** Begin Text. The paragraph itself is not further indented. It aligns with the subparagraph number.

Second Level — **(1) Title.** Begin Text. The paragraph itself is not further indented. It aligns with the subparagraph number.

Third Level — *(a)* Begin Text. The paragraph itself is not further indented. It aligns with the subparagraph number.

Fourth Level — *(1)* Begin Text. The paragraph itself is not further indented. It aligns with the subparagraph number.

* Main Subsection Headers follow the five part AASHTO format: Description, Material, Construction Requirements, Measurement, and Payment.

Figure 2-1. Hierarchal Organization of FLH Specifications

CHAPTER 3 — FORMATTING GUIDELINES

The Standard Specifications contain varying numbers and levels of subordinate sections. This format helps achieve a visually appealing document in which information is organized into a logical hierarchy that readers can use to quickly find the content they are seeking.

Table 3-1 summarizes general formatting guidelines regarding font, alignment, indentation, and spacing for the various hierarchal levels used in FLH specifications. These guidelines are based on the styles used in the FLH Supplemental Specifications and by the FLH Divisions in their Library of Specifications (LOS).

Table 3-1
Formatting Guidelines for FLH Specifications

Example	Font	Alignment	Indentation	Spacing
Section ###. — TITLE	Times New Roman 14 pt Bold	Centered	None	Before: 12 pt After: 24 pt
Subsection Header (e.g., Construction Requirements)	Times New Roman 12 pt Bold	Centered	None	Before: 12 pt After: 12 pt
XXX.XX Title. Text.	Times New Roman 12 pt Bold numbering and subsection title only	Justified	None	After: 12 pt Line Spacing: at least 12 pt
(a) Title. Text.	Times New Roman 12 pt Bold numbering and subsection title only	Justified	Left: 0.25"	After: 12 pt Line Spacing: at least 12 pt
(1) Title. Text	Times New Roman 12 pt Bold numbering and subsection title only	Justified	Left: 0.5"	After: 8 pt
(a) Text.	Times New Roman 12 pt Italicized lettering, not bold.	Justified	Left: 0.75"	After: 8 pt
(1) Text.	Times New Roman 12 pt Italicized numbering, not bold.	Justified	Left: 1.0"	After: 8 pt

Note that the FP is also set with top and bottom margins of 0.8 inches, and inside and outside (mirror) margins of 0.5 inches.

CHAPTER 4 — WORDING OF SPECIFICATIONS

4.1   VOICE AND MOOD

A specification's goal is to be *specific*. Constructing sentences using the active voice and imperative mood is the most efficient way to give a command, direction, or instruction, but is not appropriate for every situation. In writing specifications for FLH, follow these guidelines:

1. Use the active voice and imperative mood to convey instructions to the contractor. This style is most appropriate for conveying contractor responsibilities in the Material, Construction Requirements, and Measurement subsections of an FLH specification.

    Examples (Instructions to Contractor):

    Scarify gravel roads to a minimum depth of 6 inches.

    Clear the area of vegetation and obstructions according to Sections 201 and 203.

2. Use the active voice and indicative mood when it is necessary to clarify the party responsible for the action. This can occur when both Government/CO and Contractor responsibilities are discussed in the same sentence and for optional or alternative actions on the part of either the Contractor or Government (that is, discretionary clauses using *may*).

    Examples (Active Voice to Clarify Responsible Party):

    The Government and the Contractor will agree to the negotiated price.

    The CO may order the performance of the work to be stopped.

3. When stating a fact as opposed to directing an action, the indicative mood is most appropriate. The Description subsections of an FLH Standard Specification are typically written in indicative mood.

    Examples (Indicative Mood to State Facts):

    This work consists of constructing mechanically-stabilized earth walls.

    The plans indicate limits of disturbance.

4.2   WORDING

The most accurate, direct way to state a requirement is affirmatively. Positive sentences are typically shorter and easier to understand than their negative counterparts — People typically prefer to be told what to do, instead of what they cannot do.

For example,

Do not use material from a source that is unacceptable to the Government.

could be stated more simply as,

Only use material from approved sources.

Similarly, phrases are wordy when they can be replaced with fewer words that convey the same meaning. Needless words add clutter and can hinder a reader's ability to grasp what is important.

Table 4-1 suggests some alternatives to common wordy phrases and negative words.

Table 4-1
Alternatives to Common Wordy Phrases or Negative Words

Instead of:	Consider:	Instead of:	Consider:
a minimum of	at least	in lieu of	instead of
a number of	some	in many cases	often
absolutely essential	essential	in many instances	sometimes
aforementioned	the, that, those	in order to	to
as concerned with	concerns	in the amount of	for
as may be necessary	as needed	in the event of	if
as stated in	states	in the event that	if, when
at a later date	later	in the near future	soon
at the option of the contractor	the contractor may	in such a manner as to	so as to
at the present time	now	initiate	start
by means of	by	is applicable to	applies to
capability	can	is hereby authorized	may
cease and desist	stop	is indicative of	shows
commence	start	make payment	pay
consequently	so	make preparations for	prepare for
contract requirement	contract	make use of	use
cost thereof	cost of	methodology	method, way
does not have	lacks	not able	unable
does not include	excludes, omits	not accept	reject
due to the fact that	because	not certain	uncertain
enclosed herewith	enclosed	not less than	at least
endeavor	try	not many	few
for a period of	for	not often	rarely

Table 4-1
Alternatives to Common Wordy Phrases or Negative Words

Instead of:	Consider:	Instead of:	Consider:
for the purpose of	for, to	not unlike	similar, lacks
free from	without	not the same	different
give consideration to	consider	not...unless	only if
give recognition to	recognize	not...except	only if
heretofore	until now	not...until	only when
however	but	on a quarterly basis	quarterly
if the contractor so elects	the contractor may	on a regular basis	regularly
Impracticable	impractical	practicable	practical
in a manner such that	so that	prior to	before
in a timely manner	promptly, on time	subsequent to	after
in accordance with	according to	successfully complete	complete
in advance of	before	terminate	end
in an effort to	to	the month of June	June

4.3   WORDS AND PHRASES NOT TO USE

Many of the words or phrases in Table 4-2 confuse readers, do not add meaning, or introduce passages that are unnecessary because the same information is covered elsewhere in the specifications (for example, in the General Requirements).

Table 4-2
Words and Phrases to Avoid

Aforesaid	hereinbefore
and/or	in a workmanlike manner
as per	in accordance with these specifications
at the Contractor's expense	latter
authorize and direct	means and includes
care shall be taken	necessary and desirable
Department (use *Government*)	neither...nor
Engineer (use *Contracting Officer or CO*)	order and direct
entirely	pertinent
etc.	shall function as intended
former	special attention of the Contractor
full and complete	subsidiary
herein	the attention of the Contractor is directed to...
hereinafter	

4.4   SPECIFIC WORDS OR PHRASES

Use the following words and phrases in the appropriate context.

Accept vs. Approve.   In a document with legal consequences, such as the Standard Specifications, *accept* and *approve* have a difference in meaning that is important to recognize and preserve.

To *accept* is to recognize an obligation to pay, and is used in the context of, or in reference to, contracts.   To avoid misunderstanding, reserve *accept* and related forms, such as *acceptance* and *acceptable*, for use in reference to the contract between the Government and the Contractor.

In contrast, to *approve* is to confirm agreement with, or to indicate satisfaction with, a situation or circumstance.   Use *approve* and related forms, such as *approval*, to indicate official sanction or endorsement of designs, documents, plans, or processes.

All vs. Any.   *Any* and *all* should not be used interchangeably.   *All* refers to the entire amount, whereas a*ny* is a limited number selected at the discretion of the reader.   In most situations involving specified requirements, *all* is the more appropriate word.

Restrict the use of *any* to those logical situations in which meeting one criterion among several is enough to satisfy a condition.

Amount vs. Quantity.   Use *amount* when money is the subject.   Use *quantity* when volume, mass, or other unit of measurement is the subject.

And/Or.   This construction is both awkward and confusing.   Write *A, B, or both*, not *A and/or B*.

Appropriate vs. Pertinent.   Use *appropriate* (instead of *pertinent*) for stating or attaching relevant information.

As approved by the CO.   Often this phrase is not necessary as the General Requirements have already established the CO's authority over the job.   However, a variant of the phrase – *The Contractor shall obtain the CO's approval before* – is often quite useful to ensure that the Contractor consults with the CO at a critical decision point or before proceeding from one stage to another in a multi-step process.

At no additional cost to the Government.   Use *at no additional cost to the Government* instead of *at the Contractor's expense*.   The Government cannot insist that the Contractor pay for something (because the Contractor might well turn to another source to cover a cost), but it can indicate that *the Government will not pay*.

Bidder vs. Contractor.   *Bidder* is reserved for use in General Requirements related to bid procedures, the Notice to Bidders, press releases, amendments, and other similar purposes. In general, use *Contractor* in the specifications.

Contracting Officer vs. Engineer. Refer to the *Contracting Officer* or *CO*, not to the *Engineer*. The Federal Acquisition Regulations define the CO as including all the CO's representatives.

Each vs. Either. Use *either* only when a choice is implied; otherwise, use *each*.

Ensure vs. Insure vs. Assure. These are three different verbs with three different meanings. The correct word in specifications will almost always be *ensure*, which means "to make sure."

Only use *insure* when speaking of financial protection of the sort offered by insurance companies. Misusing *insure* can create or suggest an obligation vastly different from that which is intended.

Use *assure* only when giving reassurance to another person. *Assure* will rarely be the right word in a specification.

May. Use *may* when either the Contractor or Government is the subject and either or both have options or alternatives. Use *may* instead of *exercise its option to*, *reserve the right to*, or similar phrases that simply describe a party's choice or prerogatives.

Per. Use *per* when describing a rate or ratio. To avoid confusion, do not use *per* in the sense of *according to*.

Plans vs. Drawings. The FP makes a distinction between these terms when dealing with graphical design content. *Plans* are prepared by the Government and *drawings* are prepared by the Contractor.

However, based on context, *plan* could also refer to a document provided by the Contractor to describe a particular program (for example, a blasting plan, quality control plan, erosion and sedimentation control plan).

Provide vs. Furnish. Though similar, these words are not identical in meaning. *Provide* has a broader meaning, which is "to supply or make available." In contrast, *furnish* means "to equip."

Use *provide* when requiring a contractor to supply an item; because this is usually the intention in a specification, *provide* is usually the better choice of the two words.

When the intention is to additionally require that a contractor not only provide an item but also do something with it, couple *provide* with such additional verbs as *use*, *place*, or *install*.

Provide and Place vs. Construct. *Provide (or furnish) and place* should generally be reserved for items that are prefabricated. *Construct* should be used for items that are built or assembled in the field.

Section vs. Subsection. Use *Section* when referring to a section within the FP in its entirety (for example, Section 109). Use *Subsection* to refer to specific clauses (for example, Subsection 109.02) within a section.

Shall vs. Will. The term *shall* indicates an obligation to act and is reserved for Contractor responsibilities. (Use the imperative mood, active voice to avoid the use of *shall*.)

The term *will* is used to indicate an anticipated future action or result and is reserved for actions and responsibilities of the Government and Contracting Officer.

That vs. Which. Do not use *that* and *which* interchangeably. *That* is properly used to introduce information essential to the meaning of a sentence. *Which* introduces nonessential information. *That* will be the right word choice in a specification more often than *which*, for the simple reason that specifications express essential requirements.

Use the following rules to decide if a clause should start with *that* or *which*:

- If you can drop the clause and not lose the point of the sentence, use *which*. If dropping the clause would change the meaning of the sentence, use *that*.
- A *which* clause goes inside commas, a *that* clause does not.

When vs. Where vs. If. These words are not interchangeable; a writer's precision will be improved by reserving each for its most appropriate use. *When* refers to time. *Where* refers to place. *If*, among its many uses, introduces a conditional clause or sentence.

Use *when* in discussions about time or chronology. The presence of words about time, periods of time, dates, or duration are clues that point to *when* as the appropriate choice. Another clue is that *before* or *after* can often replace *when* without changing the meaning of the sentence.

Use *where* to discuss or refer to a physical place, location, or area.

Use *if* to introduce, or as part of, an *If A, then B* sentence. Do not use *when* or *where* for this purpose.

4.5   COMPOUND WORDS, HYPHENATION, AND WORD SEPARATION

English changes over time and words that are commonly used together tend to migrate, first staying paired but separate, then finding frequent use with a linking hyphen, then joining eventually into a single word.

It can be hard to know where a word pair or phrase is in this progression. Table 4-3 shows some common combinations as they should be used in FLH specifications.

Table 4-3
Compound Words, Hyphenated Words, & Word Separation

Instead of:	Use:
air entraining	air-entraining
center line	centerline
cross section	cross-section
guard rail	guardrail
job mix	job-mix
multi-lane	multilane
pre-construction	preconstruction
right of way	right-of-way
sub-base	subbase
worksite	work site

4.6 CROSS-REFERENCES AND CITATIONS

Cross-References within the Standard Specifications

Cross-references are useful to reduce repetition and to eliminate possible conflicts and ambiguities.

Examples:

Backfill according to Subsection 209.10(b).

Follow the requirements of FAR Clause 52.214-18 Preparation of Bids — Construction.

Citations

1. Treat the titles of separately issued and handled forms, certificates, standards, and similar documents as complete publications and italicize the title. Do not use quotation marks. Use the appropriate punctuation required by the sentence overall. (If, for example, the word *Form* and the form's number accurately identify a document, the title is providing supplementary information and should be set off by commas.)

   Examples:

   Maintain a *Log of Work Related Injuries and Illnesses,* OSHA Form 300, and make it available for inspection.

   Submit written job-mix formulas with Form FHWA 1641 for approval at least 28 days before production.

2. When specifying standards or test methods, identify them by their identification number such as ASTM A 307, AASHTO T 27, AASHTO M 31M, or Federal Specification TT-P-641. Do not include the year in the identification number. A reference made to a specification, standard, or test method adopted by AASHTO, ASTM, GSA, or other recognized national technical association, refers to the approved procedures that were in effect on the date of the contract solicitation. (For example, when specifying AASHTO T 27-93 use AASHTO T 27 and drop the "-93" that indicates the specification adopted for use in 1993.) An "M" after the standard number indicates a metric specification and should be included in the reference.

Refer to a national reference standard by the name of the issuing organization, followed by a space, and then the letter and the number of the standard. Place a hard (non-breaking) space between the letter and number of the standard.

Refer to the issuing organization's website to verify that the cited standards are current.

Examples:

Determine the 7-day unconfined compressive strength of the 3 mixtures according to ASTM D 633, method A.

Determine the in-place density and moisture content according to AASHTO T 310.

CHAPTER 5 — LISTS AND TABLES

5.1   LISTS

Vertical lists are often the best way to present multiple items, conditions, options, and exceptions.

As used in the FP, lists generally follow a hierarchal scheme similar to that described in Chapter 2, the exception being the exclusion of a heading title for each list item. Otherwise, list items follow the same font, alignment, spacing, and indentation guidelines as provided in Table 3-1.

Wording

To incorporate vertical lists into a specification:

1. Use a lead-in sentence punctuated with a colon to introduce the list items and to indicate the meaning or purpose of the list. If possible, explicitly identify in the lead-in sentence whether one, more than one, or all of the items apply. For example,

   - To indicate an **OR** situation, use "…one of the following:" when only one item applies.
   - To indicate an **AND/OR** situation, use one of the following:
     - "…one or both of the following:" when one or two items apply in a list of two.
     - "…one or more of the following:" when more than one item can apply individually.
     - "…one or a combination of the following:" when items can be combined.
   - To indicate an **AND** situation, use "…all of the following:" to indicate that all items apply.

2. Ensure that each item in the list fits grammatically with the lead-in sentence.

3. Make list items parallel in phrasing.

Capitalization

Capitalize the first letter of the first word of each list item.

Punctuation

The lead-in phrase introducing a list should be followed by a colon.

In a series consisting of three or more simple items, separate the items with semicolons (;) and place a period at the end of the last item. Following the second to last item in the list, include the word *and* or *or* as appropriate.

Example:

Within 24 hours, furnish inspection reports to the CO that include all of the following:

(a) Summary of the inspection;

(b) Names of personnel making the inspection;

(c) Date and time of inspection;

(d) Observations made; and

(e) Corrective action necessary, action taken, and date and time of action.

If the list items are complete sentences, punctuate each item with a period.

Example:

(b) **Tabulated schedule.** All of the following apply to the tabulated schedule:

(1) For arrow diagrams, show activity beginning and ending node numbers. For precedence diagrams, list activities and show lead or lag times.

(2) Show activity durations.

(3) Show activity descriptions.

(4) Show early start and finish dates.

(5) Show late start and finish dates.

(6) Show status (critical or not).

(7) Show total float.

5.2   TABLES

Tables are an effective method of summarizing and communicating requirements. General conventions regarding tables are provided below.

Table Numbers and Titles

Each table should have a unique number. Precede the number with the word *Table*. Begin the number with the Standard Specification section number in which the table appears, followed by a dash and the sequential number of the table within the section, starting with the numeral 1. Begin renumbering with each new section.

On the line following the table number, provide a distinct table title that conveys the contents included in the table. Capitalize the first letter of significant words in the title. Do not place a period at the end of the title.

Use bold typeface and center alignment for table numbers and titles. The table number and title should precede the table itself.

Referencing Tables

In text, introduce tables by referring to their number. Introductory phrases such as "the following table" or similar terms are not necessary.

Example:

Provide the minimum compressive strengths shown in Table 213-1.

Table Layout

Center the table, number, and title horizontally on the page. Do not allow tables to exceed the margins of the paper. Note that tables containing sampling and testing requirements (see, for example, Table 204-1) are typically rotated 90 degrees to allow them to fit on a page that has portrait orientation.

Table Notes

Place notes to a table immediately after the table to which they belong. Align notes flush with the table's left edge. Notes are in 10-pt Times Roman font.

General notes apply to the table as a whole. Introduce general notes with the word *Note* set in bold and followed by a colon.

Example:

Table 259-2
Proof Test Load Schedule

Test Load Increment	Hold Time (minutes)
AL (0.05DTL max.)	Until stable
0.25DTL	Until stable
0.50DTL	Until stable
0.75DTL	Until stable
1.00DTL	Until stable
1.25DTL	Until stable
1.50DTL (maximum load)	See below

Note: AL = Alignment load; DTL = Design test load.

Notes on specific parts of the table are introduced by a number, placed within parentheses, that refers to a numeric superscript within parentheses placed in the table.

Example:

Table 213-1
Subgrade Stabilization Strengths

Stabilization Mixture	Test Procedure	Minimum Compressive Strength
Lime/Soil	ASTM D 5102	100 pounds / square inch [1]
Lime/Fly ash/Soil	ASTM C 593	400 pounds / square inch [2]
Cement/Soil	ASTM D 1633	400 pounds / square inch [2]

(1) 28-day cure.

(2) 7-day cure followed by vacuum saturation.

CHAPTER 6 — ADDITIONAL DRAFTING CONVENTIONS

6.1 SHORTENED FORMS: ABBREVIATIONS, ACRONYMS, AND SYMBOLS

Definitions

Abbreviations, acronyms, and symbols are shortened forms of longer words, names, or expressions. Each of these short forms differs from the others in formation and usage.

Abbreviations. Shortened forms of a single word or phrase, usually followed by a period and often including lowercase letters.

> *Examples*:
>
> etc., min., max.

Acronyms. Shortened forms, often initialisms, that can be pronounced as a word. Do not use periods in an acronym.

> *Examples*:
>
> AASHTO and OSHA

Symbols. Free-standing characters, letters, or signs with unique agreed-on meanings. Do not treat or punctuate symbols as if they were abbreviations. Use a space before and after a symbol; do not precede a symbol with a hyphen or follow with a period.

> *Example*:
>
> kg, m^3

General Guidelines for Shortened Forms

As a general principle, shortened forms should be used as much for the convenience of the reader as of the writer. Adherence to the guidelines presented below will help prevent shortened forms from burdening the reader.

1. Be consistent in the use of short forms. Distinct short forms used repeatedly in FLH specifications are listed in Section 101.03 of the Standard Specifications. The accepted forms of more widely used abbreviations can be found in Chapter 14, "Abbreviations," of *The Chicago Manual of Style* or Chapters 9 and 10, "Abbreviations and Letter Symbols" and "Signs and Symbols," of the *United States Government Printing Office Style Manual, 2000*.

   If a shortened form is defined in Section 101.03, such as PVC for polyvinyl chloride, it is not necessary to precede the shortened form with the complete name in the specifications.

2. Before introducing a shortened form not listed in Section 101.03 of the FP, write out the complete name or phrase at the first usage, followed immediately with the shortened form in parentheses. When the word or phrase contains common nouns and adjectives, use lowercase letters in the full words and capital letters in the short form.

3. Do not introduce a shortened form that will not be reused in the same section; instead, simply write the words out. When reintroducing a short form in another section, write out the complete name or meaning, followed by the short form in parentheses, at the first usage.

4. Do not put an abbreviation or acronym in a title unless it is a well known, universally familiar form.

5. Use the indefinite article *an* before abbreviations and acronyms that are pronounced as if they begin with a vowel. If the short form begins as though pronounced with a consonant, use *a*.

6. Form plural short forms by adding the lowercase letter *s*. Do not use an apostrophe. For example, the plural for the abbreviation for number, *No.*, is *Nos. not* No's.

6.2   STYLE FOR MEASUREMENTS

Measurements describe quantities and consist of a *numeric value* and a *unit of measure*. General conventions regarding the use of measurements in the Standard Specifications are provided below.

1. Use numerals for the value of measurement. Provide a hard space, as described in Section 8.2, between the numeral and measurement unit. A hard space prevents the number and symbol from becoming separated across lines of text.

   Example:

Correct	Incorrect
Cover the top layer of buried debris with at least 1 foot of compacted earth.	Cover the top layer of buried debris with at least *one* foot of compacted earth.

2. For units of measure provided in text (as opposed to tables), write the full word instead of using symbols or abbreviations (with the exception of temperature measurement). In tables, symbols may be used.

Example:

Correct	Incorrect
Construct at intervals not exceeding 20 feet.	Construct at intervals not exceeding 20 *ft*.
Apply the asphalt sealant to the pavement surface at a rate of 0.20 to 0.30 gallons per square yard.	Apply the asphalt sealant to the pavement surface at a rate of 0.20 to 0.30 *gal/yd²*.

3. For temperature, use the degree symbol and the abbreviation for Fahrenheit and Celsius. Provide a hard (nonbreaking) space between the numeral and symbol. Provide no space between the degree symbol and the temperature abbreviation (for example, 7 °C, 100 °F).

   Example:

   Apply asphalt at a temperature between 150 and 175 °C.

4. For angular measurement, write out the word *degree* (for example, 90-degree angle to the vertical).

   Example:

   Place elongated pipes with major axis within 5 degrees of vertical.

5. Names of basic and derived units of measurement are always lowercased even it they are derived from a personal name (for example, newton, hertz, pascal).

   Example:

   Four roller passes of a vibratory roller having a minimum dynamic force of 180 kilonewtons impact per vibration and a minimum frequency of 16 hertz.

6. Use plural forms as necessary (for example, 1 foot vs. 2 feet; 1 meter vs. 2 meters).

   Examples:

   Place bed course material in layers not exceeding 4 inches in compacted thickness.

7. To indicate dimensionality, use the word *by* not the multiplication cross symbol (x).

   Example:

   Limit drawings to a maximum size of 610 by 920 millimeters.

8. When measurements are used as adjectives, connect the numeric value and the unit of measure with a hyphen (see Section 7.7).

6.3   MEASUREMENT SYMBOLS

Measurement symbols (for example, ft, lb, in, min) are to be used in tables and figures only. To use measurement symbols properly in tables:

1. Do not follow measurement symbols with a period unless dictated by placement at the end of a sentence. Measurement symbols are not abbreviations.

2. Do not add an *s* to form a plural. The symbol remains the same whether the quantity is one or many.

    Examples:

Correct	Incorrect
2 lb	2 lbs
24 h	24 hrs

3. Type a space between the quantity and the symbol.

    Examples:

    1 lb, 2 ft, 60 °F

4. Precede symbols only with numerals, never words.

    Example:

Correct	Incorrect
2 ft	two ft

5. Do not mix symbols and names in the same expression.

    Example:

Correct	Incorrect
ft/s	feet/s
feet per second	feet/second

6. Print symbols and quantities in normal, upright (roman) type regardless of surrounding text.

    Example:

Correct	Incorrect
2 ft	*2 ft*

7. Do not use symbolic representations.

    Example:

Correct	Incorrect
2 ft	2'
6 in	6"

6.4   MATHEMATICAL AND OTHER SIGNS AND SYMBOLS

Use signs and symbols as shown in Table 6-1:

Table 6-1
Signs and Symbols

Sign or symbol	Meaning	Use in ...	
		Tables only	Text and tables
+	Plus	✓	
−	Minus	✓	
±	plus or minus		✓
=	equal to	✓	
<	less than	✓	
≤	less than or equal to	✓	
>	greater than	✓	
≥	greater than or equal to	✓	
×	multiplied by; dimensional indicator	✓	
μ	10^{-6} ("micro")	✓	
°	degree (for temperature)		✓
§	Section	✓	
—	em dash		✓
–	en dash		✓
:	ratio; proportionality		✓
$	U.S. dollar		✓
/	Per	✓	
%	Percent	✓	

Note: When using mathematical and other signs and symbols in text (as opposed to in tables or table footnotes), use words for the quantitative relationships indicated in Table 6-1.

Example:

Correct	Incorrect
When installing culvert pipe *less than or equal to* 48 inches in diameter…	When installing culvert pipe ≤ 48 inches in diameter…
When the centerline curve radius is *greater than* 500 feet...	When the centerline curve radius is > 500 feet...

Use a colon for slope notation (vertical : horizontal). For slopes flatter than 1V:1H, express the slope as the ratio of one unit vertical to a number of units horizontal. For slopes steeper than 1V:1H, express the slope as the ratio of a number of units vertical to one unit horizontal.

6.5 RANGES

A range is defined by two endpoints. The endpoints may be inside and part of the range, or outside and excluded from the range.

Whether in text or in tables, when defining a series of related ranges that together describe a complete set of possibilities, ensure that no number or measurement can fall in more than one range. That is, make the ranges mutually exclusive.

> *Example*: One large range (0 to 750 millimeters) is divided into three mutually exclusive ranges by two internal endpoints (250 and 500 millimeters) that fall into the first and second ranges, respectively, and cannot fall elsewhere.
>
> *Correct:* From 0 to 250 millimeters; over 250 to 500 millimeters; and over 500 to 750 millimeters.
>
> *Incorrect:* From 0 to 250 millimeters; 250 to 500 millimeters; and 500 to 750 millimeters.

Ranges in Text

In text, indicate a range that includes the endpoints by using the words *from* and *to*. The words *inclusive* or *minimum* and *maximum* may be added as appropriate to enhance clarity. Do not use a dash to indicate a range (–), as this can too easily be confused with a minus sign.

Indicate a range from which the endpoints are to be excluded by using the words *between* and *and*. Most such ranges will be defined by discrete physical objects or boundaries, rather than by numeric measurements.

Avoid using *between* and *and* with measurements because this wording leads to uncertainty over how close the range should approach the endpoints. *Between 25 °C and 30 °C*, for example, could mean from 26 °C to 29 °C, or it could mean from 25.1 °C to 29.9 °C — the wording is uncertain.

> *Examples*:
>
> Provide an accurate and calibrated thermometer having a range from 200 to 600°F in 5 °F graduations.
>
> Limit joint widths from 1 inch minimum to 2 inches maximum.
>
> Take roadway cross-section data between the centerline and the new slope stake location.

Ranges in Tables

As best warranted to ensure clarity, ranges may be described in tables using symbols only, words only, or with both words and symbols. However defined, do not create adjoining ranges with shared endpoints.

Table 6-2 compares the use of these three methods to divide a range from 0 to 100, inclusive, into four contiguous smaller ranges.

Table 6-2
Displaying Ranges in Tables

Symbols only	Symbols and words combined	Words only
≤ 25	≤ 25	25 or less
>25 – 50	>25 to 50	over 25 to 50
>50 – 75	>50 to 75	over 50 to 75
>75 – 100	>75 to 100	over 75 to 100

6.6   MINIMUM, MAXIMUM, MINUTES

In text, spell out the words *minimum* and *maximum* in full or use alternatives such as *at least* or *no more than*.

Examples:

The minimum number required to perform a statistical evaluation is 3.

The maximum obtainable pay factor with 3, 4, or 5 samples is 1.01.

In tables or lists, minimum and maximum values may be indicated by using their abbreviated forms (min. and max.). In some cases, the better alternative would be to use the symbol ($\geq$) to indicate a minimum and ($\leq$) to indicate a maximum. This would help prevent confusion between the abbreviation for a minimum value (min.) from the measurement symbol for minutes (min).

Example:

Ductility, 25 °C, 50 mm/min, AASHTO T 51   40 mm min.

Or better:

Ductility, 25 °C, 50 mm/min, AASHTO T 51 $\geq$ 40 mm

CHAPTER 7 — NUMERICAL INFORMATION

7.1   NUMERALS VS. WORDS

1. Use numerals for measurements, sizes, and critical or precise quantities.

    Example:

Correct	Incorrect
a depth of 5 meters	a depth of five meters
at least 10 days	at least ten days
wrapped 1½ turns	wrapped one and one-half turns

2. Use numerals when cross-referencing sections and other parts of the specifications or similar sources.

    Example:

    Federal Acquisition Regulation (FAR), Title 48, Code of Federal Regulations, Chapter 15

3. Use words for quantities or values equal to or less than ten that do not modify measurements, sizes, or other critical or precise quantities.

    Examples:

    A divided highway has two or more roadways.

    Where there are more than five thicknesses…

4. Use numerals for values greater than ten.

5. Use words for numbers at the beginning of a sentence; if a number greater than ten appears at the beginning of a sentence, reorder the sentence if possible.

    Examples:

    Eight hours of labor constitutes a full day of work.

    Thirty minutes before installation, begin preparing the material.

    Or:

    Begin preparing the material 30 minutes before installation.

6. When quantity and size are expressed together, always use words for the quantity and numerals for the size.

    Example:

    Prepare ten 150-millimeter by 300-millimeter concrete cylinders…

7. Be consistent. Within the same context, treat similarly all numbers that refer to the same category of things.

Example:

Thirty minutes before starting, and again sixty minutes later …

7.2 TIME AND DATE

1. Use the words *noon* and *midnight* to indicate twelve o'clock. Do not use the numeral 12 followed by a word or abbreviation.

Example:

Day — Each and every day shown on the calendar, beginning and ending at midnight.

2. Use numerals for clock times. Keep zeros when describing times "on the hour." Use the standard 12-hour system, with all numerals accompanied by the appropriate a.m. or p.m. designation (using lower-case letters, followed by periods); leave a space between the numeral and abbreviation but no spaces inside the abbreviation. Do not use the abbreviation *o'clock*.

Examples:

9:00 a.m., 10:30 p.m

3. Use words (written in full) for the names of months and numerals for days of the month and years. Do not use ordinal designators (for example, th and rd) in dates.

Examples:

on June 15

from May 1 to September 30

4. Use numerals with an ordinal designator to specify a fixed number of days from an event or starting point.

Examples:

on the 15th day following receipt

the 21st day of the month

7.3 MONEY

Use numerals for monetary amounts. Use commas according to Subsection 7.8. Do not include the decimal and zeros for cents when amounts are in whole dollars; do not leave a space between the dollar sign ($) and the numeric value.

Examples:

The Government's share will not exceed $5,000.

No progress payment will be made in a month in which the work accomplished results in a net payment of less than $1,000.

7.4   DECIMALS

1. Express decimals in numerals, not words.

    Example:

Correct	Incorrect
0.1	one-tenth

2. A decimal should always have numerals on both sides.

    Example:

Correct	Incorrect
0.1	.1

3. Use decimals, not fractions, in metric expressions.

    Example:

Correct	Incorrect
8.5 kg	8½ kg

7.5   FRACTIONS

1. Use numerals for mixed numbers; do not leave a space between the whole number and fraction.

    Example:

    Over 1 to 1½

    Some fractions are available through word processing software. Format other fractions by using the superscript/subscript commands and the fraction slash..

    Examples:

    $^7/_{11}$, $^{13}/_{22}$, $^{11}/_{33}$

2. Use words for simple fractions that do not describe a measurement or a precise quantity, for fractions that stand alone, and for fractions that come before the words *of a* or *of an*. Connect the numerator and denominator with a hyphen.

    Examples:

    Replace the pile cushion if it is compressed more than one-half of its original thickness or it begins to burn.

    Do not remove mortar beyond one-third the diameter of the coarse aggregate.

7.6   DECIMALS VS. FRACTIONS

1. Use decimals, not fractions, in metric expressions.

2. Follow industry convention to choose between decimals or fractions in U.S. customary units.

Example:

Correct	Incorrect
Leave the cut end at least $\frac{1}{8}$ inch above the base.	Leave the cut end at least 0.125 inch above the base.

7.7   HYPHENS AND UNIT MODIFIERS

When a numeral and measurement unit work together to describe something else (usually an object or material, like a pipe, bolt, or board), they are acting as a single word, or adjective, called a *unit modifier*.

Examples:

Scarify to a 6-inch depth.

Use a 10-foot metal straight edge...

Use the same structure in more complex unit modifiers (that is, in phrasings with multiple adjectives working together to describe a single object).

Examples:

25-millimeter-diameter rod

2-meter-long plank

7.8   COMMAS VS. SPACES

In dollar figures, use commas in expressions with four or more digits (that is, amounts greater than $999).

Examples:

$800; $1,000; $10,000; $2,000,000.

In measurements, use commas in numeric values with five or more digits (that is, quantities greater than 9999).

Examples:

temperature of 2000 °F

10,000 pounds per square inch

1000-volt megger

CHAPTER 8 — CAPITALIZATION AND PUNCTUATION

8.1   CAPITALIZATION

Capitalize the following words or categories of specific names and things:

1. ACRONYMS (see Section 6.1);

2. Titles of documents and forms;

   Examples:

   Inspector's Daily Record of Construction Operations (Form FHWA 1413)

   OSHA Form 300

3. Laws and legislative acts;

   Example:

   Clean Air Act

4. References to divisions, sections, subsections, tables, and figures in the Standard Specifications;

   Examples:

   … as provided in Subsection 106.07

   … as shown in Tables 401-1 and 703-12

5. Proper nouns; and the

6. Common nouns identified below:

   - Contractor;
   - Government;
   - Contracting Officer or CO; and
   - Division, Section, Subsection (when referring to a numbered portion of the Standard Specifications).

Do not capitalize the following terms unless required by sentence structure or formatting conventions (for example, the first word in a sentence, first item in a list, subsection header):

- contract;
- inspector;
- state;
- agency;
- fabricator; and
- manufacturer.

8.2 PUNCTUATION

Serial Commas

In a series of three or more elements, separate the elements with a comma. Use a comma before the conjunction (*and* or *or*) joining the last two elements of the series.

Examples:

Shoring, bracing, and cofferdams will be evaluated under Subsections 106.02 and 106.04.

This work consists of processing and incorporating lime, lime/fly ash, or hydraulic cement into the upper layer of a subgrade.

Closing Quotation Marks

Place periods and commas required by a sentence *inside* closing quotation marks, regardless of whether the period or comma is part of the quoted matter.

Examples:

Use strands having similar properties, from the same source, and having the same "twist" or "lay."

Carefully pack and adequately ventilate plants to prevent "sweating."

The measured torque at a tension "P," after exceeding the turn test tension required in.......

Letters as Shapes

Type letters used as shapes in regular Times New Roman font. Do not use quotes around the letter. Link the letter and following word with a non-breaking hyphen.

Examples:

H-pile, O-ring, U-bolts, U-shaped staples

Parentheses

Use parentheses to insert and set off additional information relevant to the sentence.

Examples:

Any material misstatement by the surety, overstatement of assets (either as to ownership or value) or understatement of liabilities is cause for rejection of the surety.

A lot containing an unsatisfactory percentage of non-specification material (less than 1.00 pay factor) is accepted provided the lowest single pay factor has not fallen into the reject portion of Table 106-2.

Parentheses can be particularly useful to show related information in long or complicated sentences, or in sentences with many commas or other punctuation.

> *Example*:
>
> Drawings include, but are not limited to, layouts that show the relative position (vertical and horizontal as appropriate) of work to be performed, fabrication details for manufactured items and assemblies, installation and erection procedures...

Place commas, semicolons, periods, or other punctuation that the main sentence might need *after* the closing parenthesis mark. Do not use brackets — [ ] — or french brackets — { } — in place of parentheses; reserve brackets for mathematical formulas and equations.

Hyphens and Dashes

Although they appear relatively similar, the following are four distinct typographic characters, each with its own uses:

> hyphen: -   en dash: –   em dash: —   minus sign: –

Hyphen

The shortest of the four characters is the hyphen, which is produced directly from the keyboard. Use the hyphen to hyphenate two words in a compound adjective or words with a hyphenated prefix. Do not use a hyphen to indicate a numeric range, to connect a measurement symbol with a numeral, or as a minus sign.

> *Examples*:
>
> two-way
>
> pneumatic-tired rollers
>
> 12-inch layer

En Dash

Use the en dash in tables to indicate a numeric range (in MSWord select Insert, Symbol, Special Characters).

Em Dash

Use the em dash in section headers, in definitions, and in tables to designate empty cells (in MSWord select Insert, Symbol, Special Characters).

> *Examples*:
>
> Section 101. — TERMS, FORMAT, AND DEFINITIONS
>
> **Award** — The written acceptance of a bid by the CO.

Minus Sign

Use the minus sign in mathematical formulas and with numerals to show negative values.

Example:

temperature range of from –40 to 74 °C

Hard Spaces

Use a nonbreaking (hard) space to hold together parts of a measurement, dimension, or phrase that could cause confusion if allowed to separate at a line break (in MSWord select Insert, Symbol, Special Characters).

Use a hard space:

- between a numeral and an accompanying word (for example, July 4);
- between numerals and units (for example, 5 meters);
- between numerals and the word *percent* (for example, 90 percent)
- between the words *Section* and *Subsection* and an accompanying number (for example, Section 201); and
- between the letter and the number of an ASTM and similar specifications (for example, ASTM C 595).

Example:

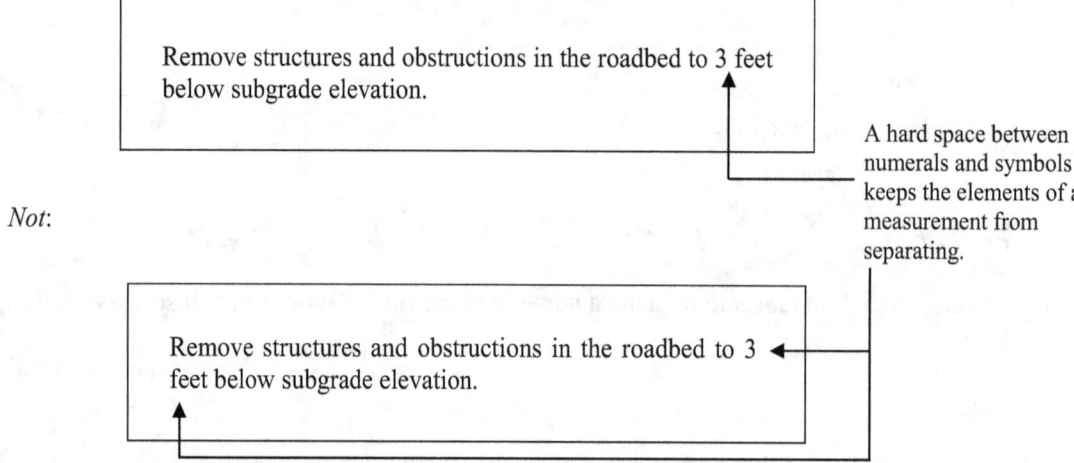

Remove structures and obstructions in the roadbed to 3 feet below subgrade elevation.

A hard space between numerals and symbols keeps the elements of a measurement from separating.

Not:

Remove structures and obstructions in the roadbed to 3 feet below subgrade elevation.

8.3   ITALICS

Use italics (characters set in type that slants to the right *like this*) as opposed to roman type in the following circumstances:

1. Use italics to denote certain hierarchal levels in specifications as identified in Table 3-1.

2. Use italics to cite complete titles of books, forms, standards, and similar documents as discussed in Section 4.6. Note that italics do not include punctuation marks (end marks or parentheses, for instance) next to the words being italicized unless those punctuation marks are meant to be considered as part of what is being italicized.

3. Use italics, as opposed to quotation marks, to refer to words that are being talked about.

   Example:

   Wherever *directed*, *required*, *prescribed*, or other similar words are used, the *direction*, *requirement*, or *order* of the CO is intended.

4. Use italics to identify mathematical symbols used in equations.

   Example:

   Calculate the upper quality index (Q_U): $Q_U = \dfrac{USL - \overline{X}}{s}$

   where: USL = upper specification limit

5. Although not a common application of italics in specifications, italics may also be used for emphasis or contrast, that is, to distinguish certain words from others within the text.

   Example:

   Wherever *"directed,"* *"required,"* *"prescribed,"* or other similar words are used, the *"direction,"* *"requirement,"* or *"order"* of the Contracting Officer is intended. Similarly, wherever *"approved,"* *"acceptable,"* *"suitable,"* *"satisfactory,"* or similar words are used, the words mean *"approved by,"* *"acceptable to,"* or *"satisfactory to"* the Contracting Officer.

APPENDIX A

A.1   SUBMITTAL EXAMPLE FOR PROPOSED SUPPLEMENTAL SPECIFICATION

<u>When slurry seals or micro-surfacing is required, include the following:</u>   (6/4/07)

Delete the first paragraph of Subsection 410.03 and substitute the following:

410.03 Composition of Mix (Job-Mix Formula).   Furnish a slurry seal or micro-surfacing mixture of aggregate, water, emulsified asphalt, or polymer modified asphalt and additives according to ASTM D 3910, ISSA A 105, and ISSA A 143.   Conform to the applicable aggregate gradation in Table 703-8 and the residual asphalt contents in Subsection 410.01.

{REASON: In response to an 4/6/06 format question from Wade Johnson, Brad Neitzke reviewed the specification references in Subsection 410.03 and determined that the reference to ISSA T 114 should be changed to ISSA A 105 and A 143 as described in his 4/10/06 e-mail. This change implements Brad's recommendation.}

(FP-03 Metric version, p. 277)

www.ingramcontent.com/pod-product-compliance
Lightning Source LLC
Chambersburg PA
CBHW081237170526
45165CB00009B/3088